Infinity and Our Unbounded Universe

The concept of Infinity has been around since the ancients and it has many practical applications in mathematics like Calculus.

Infinity is also important in Fractals which are a graphical algorithm which is used in many computer graphics applications today and which has infinite depth.

It is also applicable to our Universe since the standard concept we are taught about the Big Bang as the beginning of the Universe may be wrong.

Modern evidence is showing us contradictions which say that we live in a Universe which is steady state and might even be infinite in size. The Universe doesn't have a creation date we can be sure of.

We don't really know if it is infinite in size but the Universe is certainly much larger and older than our best scientific instruments can measure.

This book is an exploration of Infinity and new understandings of our Universe which have come to light in recent decades.

Infinity and Our Unbounded Universe

Infinity and Our Unbounded Universe

Copyright Page

This book is copyrighted for 2022

Title: Infinity and our Unbounded Universe

Subtitle: The Big Bang Never Happened

The Crazy and Out of the Box Series Book 13

By Martin K. Ettington

All Rights Reserved USA 2022

ISBN: 9798358988552

Printed in the United States of America

Infinity and Our Unbounded Universe

Infinity and Our Unbounded Universe

Other books by Martin K. Ettington

Spiritual and Metaphysics Books:
Prophecy: A History and How to Guide
God Like Powers and Abilities
Enlightenment for Newbies
Removing Illusions to Find True Happiness
Using the Scientific Method to Study the Paranormal
A Compendium of Metaphysics and How to Guides (Six books together in one volume)
Love from the Heart
The Enlightenment Experience
Learn Your Soul's Purpose
Pursuing Enlightenment
A Modern Man's Search for Truth
Use Intuition and Prophecy to Improve Your Life
The Handbook of Spiritual and Energy Healing
Pure Spirituality and God
Memories Before Birth and Reincarnation
Paranormal Abilities and the Yoga Sutras of Patanjali
Mystical and Magical Societies and Practitioners
Important Prophecies of the Future

Longevity & Immortality:
Physical Immortality: A History and How to Guide
The Commentaries of Living Immortals

Records of Extremely Long Lived Persons
Enlightenment and Immortality
Longevity Improvements from Science
The 10 Principles of Personal Longevity
Telomeres & Longevity
The Diets and Lifestyles of the World's Oldest Peoples
The Longevity Six Books Bundle
Long Lived Plants and Animals
A Guide to Longevity Foods, Diets, and Supplements

Science Fiction:
Out of This Universe

The Immortals of the Interstellar Colony
The Psychic Soldier
The Immortality Sci Fi Bundle
Visiting Many Universes

The History of Science Fiction and Fantasy

The God Like Powers Series:
Human Invisibility
Invulnerability and Shielding
Teleportation
Psychokinesis
Our Energy Body, Auras, and Thoughtforms
The God Like Powers Series— Volume 1 Compilation

The Yoga Discovery Series:
Yoga-An Ancient Art Form
Hatha Yoga-Helping you Live Better
Raja Yoga-Through the Ages
The Yoga Discovery Package

Business & Coaching Books:
Creating, Paublishing, & Marketing Practitioner Ebooks
Building a Successful Longevity Coaching Business
Why Become a Coach?
The Professional Coaching Success Trilogy
2020-Make Money Writing and Selling Books
The 2020 Handbook of High Paying Work Without a College Degree
The important of Creativity and How to Improve Yours
Quantum Mechanics, Technology, Consciousness, and the Multiverse

Self-Improvement
Stress Relief and Methods to do So
The Importance of Creativity and How to Improve Yours
Building Self-Confidence
See the World Clearly
A Trilogy of Self Help Books
A New Paradigm of Truth and Happiness
Building Hope and Wonder Among Chaos
The Importance of Genius In Our World

Science, Technology, and Misc.
Future Predictions By and Engineer & Seer
The Unusual Science & Technology Bundle

Infinity and Our Unbounded Universe

Removing Limits On Our Consciousness-And Thinking Outside the Box
Universal Holistic Philosophy
Ball Lightning
Stranger Than Science Stories and Facts
Planet Earth is Conscious

Survival
Survival of Humanity Throughout the Ages
33 Incredible True Survival Stories
The Importance of Fire in History and Mythology
How to Survive Anything: From the Wilderness to Man Made Disasters
Building and Stocking a Nuclear Shelter for less than $10,000
The Human Survival Five Books Bundle
Stranger Than Science Facts and Stories
Stranger Than Science Facts and Stories Volume Two
The Microscopic World Inside and Around Us

Legendary Beings
Are Cryptozoological Animals Real or Imaginary?
Fire in History and Mythology
All About Dragons
Sea Serpents and Ocean Monsters
The Legendary Animals Five Books Bundle
The Mythical People of Ireland
Bigfoot Mysteries and Some Answers
About the Little People: Fairies, Elves, Dwarfs and Leprechauns

Ancient History
The Real Atlantis-In the Eye of the Sahara
Ancient & Prehistoric Civilizations
Ancient & Prehistoric Civilizations-Book Two
The History of Antediluvian Giants
The Antediluvian History of Earth
Ancient Underground Cities and Tunnels
Strange Objects Which Should Not Exist
More Out of Place Artifacts
Strange and Ancient Places in the USA
A Theory of Ancient Prehistory And Giant Aliens
The Destruction of Civilization About 10,500 B.C.
A Timeline of Intelligent Life on Earth
A 300 Million Year Old Civilization Existed on Earth
The Encyclopedia of Out of Place Artifacts
Hollow and Inner Earth Stories and Facts

Aliens and Space
Types of UFOs Observed in History
Aliens and Secret Technology
Aliens Are Already Among Us
Designing and Building Space Colonies
Humanity and the Universe

Living in Space
All About Moon Bases
All About Mars Journeys and Settlement
The Space and Aliens Six Books Bundle
The Space Colonies and Space Structures Coloring Book
All About Asteroids
Spaceships, Past, Present, and Future
Astronauts, Cosmonauts, and Other Important Space Flyers
All About Mars Journeys and Settlement
Mining the Asteroid Belt
The New Era of Space Stations
Moon Landings, Bases, and Exploration

Time Travel and Dimensions
Real Time Travel Stories From a Psychic Engineer
The Real Nature of Time: An Analysis of Physics, Prophecy, and Time Travel Experiences
Stories of Parallel Dimensions
We Live in a Malleable Reality-and We Can Change It
The Time, Dimensions, and Quantum Mechanical Bundle
Alternate Dimensions & the Otherworld

The Multiverse: Time and Dimensional Travel Q&As

Political and Social

The Empire of the United States: Forged By God's Spirit Through Man

Infinity and Our Unbounded Universe

<u>The Longevity Training Series</u>

(A transcription of the online Multimedia Longevity Coaching Training Program)

The Personal Longevity Training Series-Book1-Long Lived Persons
The Personal Longevity Training Series-Book2-Your Soul's Purpose
The Personal Longevity Training Series-Book3-Enable Your Life Urge
The Personal Longevity Training Series-Book4-Your Spiritual Connection
The Personal Longevity Training Series-Book5-Having Love in Your Heart
The Personal Longevity Training Series-Book6-Energy Body Health
The Personal Longevity Training Series-Book7-The Science of Longevity
The Personal Longevity Training Series-Book8-Physical Body Health
The Personal Longevity Training Series-Book9-Avoiding Accidents
The Personal Longevity Training Series-Book10-Implementing These Principles

The Personal Longevity Training Series-Books One Thru Ten

These books are all available in digital and printed formats from my website and on Amazon, Barnes & Noble, Apple ITunes, and many other sites

My Books Website is: http://mkettingtonbooks.com

Infinity and Our Unbounded Universe

Signup for our Mailing List to get the following:

1) A discount coupon for 25% discount on all books on our site
2) Occasional Notices of new books available
3) Occasional Email on other offerings of ours (Monthly)

If you have any questions about this book or other subjects please contact the Author at:

mke@mkettingtonbooks.com

Infinity and Our Unbounded Universe

Table of Content

1.0	Introduction	1
2.0	The History of Infinity	3
3.0	George Cantor	7
4.0	Fractals Have Infinite Depth	9
5.0	What are Practical Applications of Infinity?	13
6.0	Our Unbounded Universe	21
7.0	Theory of the Big Bang	23
8.0	The Big Bang Never Happened	27
8.1	Stars Older than the Universe	33
8.2	Specific Evidence Refuting the Big Bang	41
8.3	Dark Matter Doesn't Exist?	49
9.0	Implications of an Infinite Universe	57
10.0	Summary	61
11.0	Bibliography	63

Infinity and Our Unbounded Universe

Infinity and Our Unbounded Universe

1.0 Introduction

The concept of Infinity has been around since the ancients and it has many practical applications in mathematics like Calculus.

Infinity is also important in Fractals which are a graphical algorithm which is used in many computer graphics applications today and which has infinite depth.

It is also applicable to our Universe since the standard concept we are taught about the Big Bang as the beginning of the Universe may be wrong.

Modern evidence is showing us contradictions which say that we live in a Universe which is steady state and might even be infinite in size. The Universe doesn't have a creation date we can be sure of.

We don't really know if it is infinite in size but the Universe is certainly much larger and older than our best scientific instruments can measure.

This book is an exploration of Infinity and new understandings of our Universe which have come to light in recent decades.

Infinity and Our Unbounded Universe

Infinity and Our Unbounded Universe

2.0 The History of Infinity

What is Infinity? Infinity is that which is boundless, endless, or larger than any natural number. It is often denoted by the infinity symbol.

Since the time of the ancient Greeks, the philosophical nature of infinity was the subject of many discussions among philosophers. In the 17th century, with the introduction of the infinity symbol and the infinitesimal calculus, mathematicians began to work with infinite series and what some mathematicians (including l'Hôpital and Bernoulli) regarded as infinitely small quantities, but infinity continued to be associated with endless processes. As mathematicians struggled with the foundation of calculus, it remained unclear whether infinity could be considered as a number or magnitude and, if so, how this could be done. At the end of the 19th century, Georg Cantor enlarged the mathematical study of infinity by studying infinite sets and infinite numbers, showing that they can be of various sizes.

For example, if a line is viewed as the set of all of its points, their infinite number (i.e., the cardinality of the line) is larger than the number of integers. In this usage, infinity is a mathematical concept, and infinite mathematical objects can be studied, manipulated, and used just like any other mathematical object.

The mathematical concept of infinity refines and extends the old philosophical concept, in particular by introducing infinitely many different sizes of infinite sets. Among the axioms of Zermelo–Fraenkel set theory, on which most of modern mathematics can be developed, is the axiom of infinity, which guarantees the existence of infinite sets. The mathematical concept of infinity and the manipulation of infinite sets are used everywhere in mathematics, even in areas such as combinatorics that may seem to have nothing to do with them. For example, Wiles's proof of Fermat's Last Theorem implicitly relies on the existence of very large infinite sets for solving a long-standing problem that is stated in terms of elementary arithmetic.

In physics and cosmology, whether the Universe is infinite is an open question.

Early Greek

The earliest recorded idea of infinity in Greece may be that of Anaximander (c. 610 – c. 546 BC) a pre-Socratic Greek philosopher. He used the word apeiron, which means "unbounded", "indefinite", and perhaps can be translated as "infinite".

Aristotle (350 BC) distinguished potential infinity from actual infinity, which he regarded as impossible due to the various paradoxes it seemed to produce. It has been argued that, in line with this view, the Hellenistic Greeks had a "horror of the infinite" which would, for example, explain why Euclid (c. 300 BC) did not say that there are an infinity of primes but rather "Prime numbers are more than any assigned multitude of prime numbers." It has also been maintained, that, in proving the infinitude of the prime numbers, Euclid "was the first to overcome the horror of the infinite". There is a similar controversy concerning Euclid's parallel postulate, sometimes translated:

If a straight line falling across two [other] straight lines makes internal angles on the same side [of itself whose sum is] less than two right angles, then the two [other] straight lines, being produced to infinity, meet on that side [of the original straight line] that the [sum of the internal angles] is less than two right angles.

Other translators, however, prefer the translation "the two straight lines, if produced indefinitely ...", thus avoiding the implication that Euclid was comfortable with the notion of infinity. Finally, it has been maintained that a reflection on infinity, far from eliciting a "horror of the infinite", underlay all of early Greek philosophy and that Aristotle's "potential infinity" is an aberration from the general trend of this period.

Infinity and Our Unbounded Universe

Infinity and Our Unbounded Universe

3.0 George Cantor

Georg Ferdinand Ludwig Philipp Cantor March 3 1845 – January 6, 1918 was a German mathematician. He was the most influential mathematician in the study of infinity.

He played a pivotal role in the creation of set theory, which has become a fundamental theory in mathematics. Cantor established the importance of one-to-one correspondence between the members of two sets, defined infinite and well-ordered sets, and proved that the real numbers are more numerous than the natural numbers. In fact, Cantor's method of proof of this theorem implies the existence of an infinity of infinities. He defined the cardinal and ordinal numbers and their arithmetic. Cantor's work is of great philosophical interest, a fact he was well aware of.

Originally, Cantor's theory of transfinite numbers was regarded as counter-intuitive – even shocking. This caused it to encounter resistance from mathematical contemporaries such as Leopold Kronecker and Henri Poincaré and later from Hermann Weyl and L. E. J. Brouwer, while Ludwig Wittgenstein raised philosophical objections.

Infinity and Our Unbounded Universe

Cantor, a devout Lutheran Christian, believed the theory had been communicated to him by God. Some Christian theologians (particularly neo-Scholastics) saw Cantor's work as a challenge to the uniqueness of the absolute infinity in the nature of God – on one occasion equating the theory of transfinite numbers with pantheism – a proposition that Cantor vigorously rejected. It is important to note that not all theologians were against Cantor's theory; prominent neo-scholastic philosopher Constantin Gutberlet was in favor of it and Cardinal Johann Baptist Franzelin accepted it as a valid theory (after Cantor made some important clarifications).

The objections to Cantor's work were occasionally fierce: Leopold Kronecker's public opposition and personal attacks included describing Cantor as a "scientific charlatan", a "renegade" and a "corrupter of youth". Kronecker objected to Cantor's proofs that the algebraic numbers are countable, and that the transcendental numbers are uncountable, results now included in a standard mathematics curriculum. Writing decades after Cantor's death, Wittgenstein lamented that mathematics is "ridden through and through with the pernicious idioms of set theory", which he dismissed as "utter nonsense" that is "laughable" and "wrong". Cantor's recurring bouts of depression from 1884 to the end of his life have been blamed on the hostile attitude of many of his contemporaries, though some have explained these episodes as probable manifestations of a bipolar disorder.

The harsh criticism has been matched by later accolades. In 1904, the Royal Society awarded Cantor its Sylvester Medal, the highest honor it can confer for work in mathematics. David Hilbert defended it from its critics by declaring, "No one shall expel us from the paradise that Cantor has created."

Infinity and Our Unbounded Universe

4.0 Fractals Have Infinite Depth

Fractals have been developed over the last few decades and are now used in many computer animation programs.

In mathematics, a fractal is a geometric shape containing detailed structure at arbitrarily small scales, usually having a fractal dimension strictly exceeding the topological dimension. Many fractals appear similar at various scales, as illustrated in successive magnifications of the Mandelbrot set. This exhibition of similar patterns at increasingly smaller scales is called self-similarity, also known as expanding symmetry or unfolding symmetry; if this replication is exactly the same at every scale, as in the Menger sponge, the shape is called affine self-similar. Fractal geometry lies within the mathematical branch of measure theory.

Infinity and Our Unbounded Universe

One way that fractals are different from finite geometric figures is how they scale. Doubling the edge lengths of a filled polygon multiplies its area by four, which is two (the ratio of the new to the old side length) raised to the power of two (the conventional dimension of the filled polygon).

Likewise, if the radius of a filled sphere is doubled, its volume scales by eight, which is two (the ratio of the new to the old radius) to the power of three (the conventional dimension of the filled sphere). However, if a fractal's one-dimensional lengths are all doubled, the spatial content of the fractal scales by a power that is not necessarily an integer and is in general greater than its conventional dimension. This power is called the fractal dimension of the geometric object, to distinguish it from the conventional dimension (which is formally called the topological dimension).

Analytically, many fractals are nowhere differentiable. An infinite fractal curve can be conceived of as winding through space differently from an ordinary line – although it is still topologically 1-dimensional, its fractal dimension indicates that it locally fills space more efficiently than an ordinary line.

A line segment is similar to a proper part of itself, but hardly a fractal. Starting in the 17th century with notions of recursion, fractals have moved through increasingly rigorous mathematical treatment to the study of continuous but not differentiable functions in the 19th century by the seminal work of Bernard Bolzano, Bernhard Riemann, and Karl Weierstrass, and on to the coining of the word fractal in the 20th century with a subsequent burgeoning of interest in fractals and computer-based modelling in the 20th century.

Infinity and Our Unbounded Universe

There is some disagreement among mathematicians about how the concept of a fractal should be formally defined. Mandelbrot himself summarized it as "beautiful, damn hard, increasingly useful. That's fractals."

More formally, in 1982 Mandelbrot defined fractal as follows: "A fractal is by definition a set for which the Hausdorff–Besicovitch dimension strictly exceeds the topological dimension." Later, seeing this as too restrictive, he simplified and expanded the definition to this: "A fractal is a rough or fragmented geometric shape that can be split into parts, each of which is (at least approximately) a reduced-size copy of the whole." Still later, Mandelbrot proposed "to use fractal without a pedantic definition, to use fractal dimension as a generic term applicable to all the variants".

The consensus among mathematicians is that theoretical fractals are infinitely self-similar iterated and detailed mathematical constructs, of which many examples have been formulated and studied. Fractals are not limited to geometric patterns, but can also describe processes in time. Fractal patterns with various degrees of self-similarity have been rendered or studied in visual, physical, and aural media and found in nature, technology, art, architecture and law. Fractals are of particular relevance in the field of chaos theory because they show up in the geometric depictions of most chaotic processes (typically either as attractors or as boundaries between basins of attraction).

Infinity and Our Unbounded Universe

Infinity and Our Unbounded Universe

5.0 What are Practical Applications of Infinity?

The concept of infinity is frequently used in mathematics and especially in Calculus which is a heavily used tool in Engineering and many scientific analysis techniques.

Here are some ways that infinity is used in mathematics. We start with Limits:

Limits can be written in the form

$$\lim_{x \to c} f(x) = k$$

where c and k are constants, and f(x) is a function. You can, however, have limits that are *evaluated at infinity* or have an *evaluated value of infinity*. While infinity is a strange concept, we can use it to determine the behavior of functions. This leads us to the discussion of infinite limits and limits at infinity.

Infinite limits

Infinite limits are limits that evaluate to infinity. They look something like:

$$\lim_{x \to c} f(x) = \infty \quad \text{or} \quad \lim_{x \to c} f(x) = -\infty$$

Basically, what this says is that as x approaches some x-value, c, the function f(x) blows up in the positive direction (given by the positive infinity) or blows up in the negative direction (given by the negative infinity). In other words, the function is getting super large in the positive or negative direction without bound.

Infinity and Our Unbounded Universe

While infinity is not technically a "value" that f(x) is approaching, we can imagine it as such. In general, if a limit evaluates to infinity, the limit would not exist; however, by writing it out as

$$\lim_{x \to c} f(x) = \infty \qquad \text{or} \qquad \lim_{x \to c} f(x) = -\infty$$

Instead of merely stating "the limit does not exist," we provide a lot more information about the behavior of the function at a certain value.

Usually, you will see one-sided limits when dealing with infinite limits. This is because infinite limits generally imply some kind of *vertical asymptote*, which are described separately from the left and from the right. Asymptotes are a type of discontinuity and are very important for analyzing the behavior of functions.

Here's how you can tell what type of behavior you have at some x-value, c, of a function using one-sided infinite limits. Note that each of the following shows asymptotic behavior.

Infinity and Our Unbounded Universe

Infinite limit	Behavior
$\lim_{x \to c^+} f(x) = \infty$	f(x) approaches positive infinity from the right of c
$\lim_{x \to c^-} f(x) = \infty$	f(x) approaches positive infinity from the left of c
$\lim_{x \to c^+} f(x) = -\infty$	f(x) approaches negative infinity from the right of c
$\lim_{x \to c^-} f(x) = -\infty$	f(x) approaches negative infinity from the left of c

If you're wondering just how a limit can evaluate to infinity (or in other words, how we can create an asymptote) just take a look at the function f(x) = 1/x. Imagine what would happen to the graph of the function as x got really close to 0. That's right – the denominator would get super small, thus making the fraction 1/x super large. As you take smaller and smaller values of x close to 0, the values of f(x) will grow quickly without bound, generating an asymptote. Of course, whether the values go to positive or negative infinity depends on which side we are

Infinity and Our Unbounded Universe

approaching 0 from, which shows why one-sided limits are extremely important when dealing with infinite limits.

As you can see, when we have a denominator get super close to 0, we can (but not necessarily) have an infinite limit, for our fraction gets super large without bound. Note that asymptotes apply mostly to rational functions (whose numerator and denominator are polynomials with degree greater than 0), and *do not* show up in standard polynomials. This is because polynomials don't have anything that causes them to "blow up" to infinity or negative infinity, but many rational functions have denominators that do.

There is a general procedure for finding asymptotes for rational functions:

factor the numerator
factor the denominator
cancel any common factors in the numerator and the denominator
set the denominator to 0

The reason we factor the numerator is because we want to make sure nothing cancels. For instance, if we had an (x-3) term in the numerator of a rational function and a (x-3) in the denominator as well, then these would cancel, creating a "hole" at x =3, but there would be no asymptote! Next, we must factor the denominator and set each factor equal to 0 to find the x-values that will make the denominator 0. As x approaches any of these values, the denominator will get infinitely small and the rational function infinitely large.

Infinity and Our Unbounded Universe

Limits at infinity

Don't be fooled – limits at infinity are not the same as infinite limits. While they both are some kind of limit and have something to do with infinity, they have different meanings and applications.

While infinite limits evaluate to infinity, limits at infinity are evaluated *at* infinity. Now this might sound strange, because infinity is not technically a "value" at which you can evaluate a limit. However, we can treat limits at infinity as an analysis of a function's *end behavior*. In other words, it answers the question "As the values of x get larger (either in the positive or negative directions), what happens to the y-values of the function?"

Note that we are not actually evaluating our function at infinity but rather thinking about our function as x approaches larger positive or negative values.

When we say limit at infinity, we mean one of two things:

$$\lim_{x \to \infty} f(x) \qquad \text{or} \qquad \lim_{x \to -\infty} f(x)$$

The limit on the left is looking at f(x) as x gets larger and larger in the positive direction, thus analyzing the right-ended behavior of the function. On the other hand, the limit on the right analyzes the left-handed behavior of the function, as it is looking at f(x) as x gets larger in the negative direction.

So, what is the purpose of evaluating a limit at infinity? What does it tell us exactly? As we mentioned earlier, limits at infinity help us look at the end behavior of

Infinity and Our Unbounded Universe

functions. The end behavior of a function f(x) is the behavior of its graph as the values of x approach infinity and negative infinity. You may have heard the term "end behavior" used in previous math courses when looking at polynomials, and you were probably told to look at the leading coefficient and degree to determine end behavior. What you did then was essentially using limits. Let's take a look at an example why.

Determine the end behavior of $P(x) = 3x^4 + 3x^3 + 2$. Before, you would probably look at this and say "The polynomial has an even degree, namely 4, and the leading coefficient is positive; so, the graph of P(x) will approach infinity as x approaches both infinity and negative infinity." This is very similar to how you would evaluate end behavior using limits.

Say we took:

$$\lim_{x \to \infty} P(x)$$

Unfortunately, we can't literally plug infinity into P(x), so we have to think about it. What we want to know is what happens to P(x) when x gets really large in the positive direction. If you try plugging in increasingly large x-values into P(x), you will notice that it begins to grow faster and faster. This makes sense, because the term with the largest power should eventually take control of the behavior of P(x), and the other terms will have a negligible effect. Since x^4 has the largest degree, we are essentially looking at:

$$\lim_{x \to \infty} x^4$$

Infinity and Our Unbounded Universe

This clearly blows up to a really big number, which is consistent with our earlier finding that P(x) will approach infinity as x approaches infinity. We should also take a look at our leading coefficient, but since it is positive here, it won't really affect our limit. However, if we had a negative leading coefficient, our values would approach *negative infinity* instead.

Infinity and Our Unbounded Universe

Infinity and Our Unbounded Universe

6.0 Our Unbounded Universe

What if we live in a steady state or infinite Universe which is well beyond our abilities to measure it?

This is a major contradiction of an established concept and heresy for most scientists since we were all taught growing up that the Universe is finite and was created in the Big Bang about 13.7 Billion years ago.

In the following chapters we review how the Big Bang Theory came to be and explore the evidence that the Universe is actually "Open" or unending and what are the practical implications if this is true?

Infinity and Our Unbounded Universe

Infinity and Our Unbounded Universe

7.0 Theory of the Big Bang

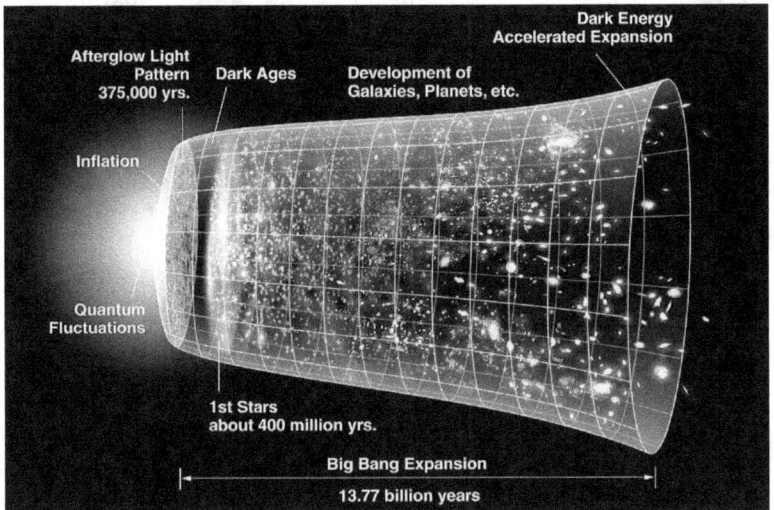

The Big Bang has been taught in schools and Universities for over one hundred years and is stated as a truth of the construction of the cosmos.

The Big Bang event is a physical theory that describes how the universe expanded from an initial state of high density and temperature. Various cosmological models of the Big Bang explain the evolution of the observable universe from the earliest known periods through its subsequent large-scale form. These models offer a comprehensive explanation for a broad range of observed phenomena, including the abundance of light elements, the cosmic microwave background (CMB) radiation, and large-scale structure. The overall uniformity of the Universe, known as the flatness problem, is explained through cosmic inflation: a sudden and very rapid expansion of space during the earliest moments. However, physics currently lacks a widely accepted theory of quantum gravity that can successfully model the earliest conditions of the Big Bang.

Infinity and Our Unbounded Universe

Crucially, these models are compatible with the Hubble–Lemaître law—the observation that the farther away a galaxy is, the faster it is moving away from Earth.

Extrapolating this cosmic expansion backwards in time using the known laws of physics, the models describe an increasingly concentrated cosmos preceded by a singularity in which space and time lose meaning (typically named "the Big Bang singularity"). In 1964 the CMB was discovered, which convinced many cosmologists that the competing steady-state model of cosmic evolution was falsified, since the Big Bang models predict a uniform background radiation caused by high temperatures and densities in the distant past. A wide range of empirical evidence strongly favors the Big Bang event, which is now essentially universally accepted. Detailed measurements of the expansion rate of the universe place the Big Bang singularity at an estimated 13.787 ± 0.020 billion years ago, which is considered the age of the universe.

There remain aspects of the observed universe that are not yet adequately explained by the Big Bang models. After its initial expansion, the universe cooled sufficiently to allow the formation of subatomic particles, and later atoms.

The unequal abundances of matter and antimatter that allowed this to occur is an unexplained effect known as baryon asymmetry. These primordial elements—mostly hydrogen, with some helium and lithium—later coalesced through gravity, forming early stars and galaxies.

Astronomers observe the gravitational effects of an unknown dark matter surrounding galaxies. Most of the gravitational potential in the universe seems to be in this form, and the Big Bang models and various observations

Infinity and Our Unbounded Universe

indicate that this excess gravitational potential is not created by baryonic matter, such as normal atoms.

Measurements of the redshifts of supernovae indicate that the expansion of the universe is accelerating, an observation attributed to an unexplained phenomenon known as dark energy.

Infinity and Our Unbounded Universe

Infinity and Our Unbounded Universe

8.0 The Big Bang Never Happened

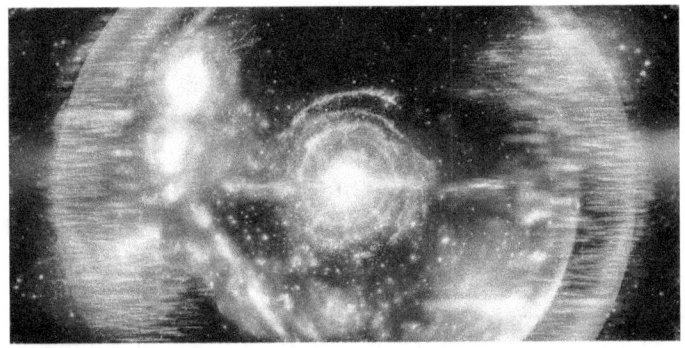

When I grew I was taught that the Big Bang Theory of the creation of the Cosmos was correct and indisputable. Now, scientists are finding out that evidence leads to conclusions that this is not true and we live in a steady state Universe. If this is true it would be an incredible paradigm change.

A tenet of the Big Bang theory, particularly that it produced conditions for certain elements to develop, is on the verge of being dramatically overturned by a scientist who claims evidence shows the event never happened, according to a study.

Under the Big Bang scenario, an explosion occurred at the dawn of our universe 13.7 billion years ago that dispersed chemical elements across space which cooled and formed the galaxies and stars in our cosmos. Modern astronomy's study of the origin and ongoing development of our universe is built largely on the dominant theory's central hypothesis.

Infinity and Our Unbounded Universe

But three critical fusion events believed to have been created by the Big Bang are under intense scrutiny by scientist Eric J. Lerner of the nuclear fusion research company LPPFusion.

Scientists believe that precise amounts of helium, deuterium and lithium were produced by fusion reactions in the dense, extremely hot cloud of chemical elements that emerged after the Big Bang.

Lerner, who has spent decades making detailed observations of such reactions, says his and other scientists' findings don't match up with longstanding theories based on observations of older stars. He found that old stars had less than half the helium and less than one tenth the lithium than is predicted by the Big Bang nucleosynthesis theory, which posits that a quarter of the universe's mass is comprised of helium.

According to Lerner – who wrote the book "The Big Bang Never Happened" – no helium or lithium was created before the development of the first stars in our galaxy.

In a statement, Lerner said the mismatch of evidence on the presence of lithium in the cosmos has been well-known for some time among astronomers. But he says challenges to the dominant Big Bang theory – such as the closed-universe and Hubble-constant problems and the failure to find evidence of dark matter – have been dismissed by scientists.

Infinity and Our Unbounded Universe

"The Big Bang should have resulted in the annihilation of matter and antimatter, leaving a surviving density of matter that would be a hundred billion times less than that observed," Lerner said in the statement. "To avoid that outcome, Big Bang theory requires an asymmetry of matter and antimatter with consequences, such as the decay of the proton, which have been contradicted by extensive experiments."

In another example, Lerner claims that in a galaxy that is expanding, as the Big Bang theory posits, the surface brightness of distant galaxies should decline over time.

"For cosmology to advance, the basic hypothesis of the Big Bang has to be abandoned," Lerner said in the statement. "The real crisis in cosmology is that the Big Bang never happened."

Lerner says the Galactic Origin of Light Elements, or GOLE, hypothesis, rightly holds that the first generation of stars to form in the cosmos were stars of intermediate mass roughly four to 12 times the size of our sun.

Under the GOLE theory, helium, deuterium and lithium were produced by these stars after they burned hydrogen at faster rates than our sun and dispersed elements across the cosmos through stellar winds.

New observations based on the GOLE hypothesis show that the early stars also produce carbon, boron and beryllium in the amounts observed in the oldest stars.

Infinity and Our Unbounded Universe

Lerner said his findings are buttressed by his recent observations of newly formed, more luminous galaxies.

"The correct predictions of the GOLE model not only fit the observations far better than does the Big Bang model" Lerner said in the statement. "The production of the light elements by stars must occur – and if there was also production by a Big Bang, we would observe far more of these light elements than we do."

Not everyone who studies space and the cosmos is ready to get on board with Lerner's theory, however. A Los Angeles-based astronomy and physics professor said longstanding scientific evidence refutes Lerner's claims.

"Many of his arguments don't hold water," University of Southern California professor Vahé Peroomian said in an interview, noting Lerner seldom links to peer-reviewed articles. "My general impression would be to take things he argues with a grain of salt."

Peroomian said that cosmic microwave background, for example, which is evidence of radiation stemming from the Big Bang, is a pillar of the cosmological theory and one that Lerner cannot dispute.

Also, if there were major flaws with the Big Bang theory, Lerner wouldn't be the only critical voice rising from the scientific community, Peroomian said.

Infinity and Our Unbounded Universe

Peroomian – who is not a cosmologist – pointed to astrophysicist Edward L. Wright's extensive critique of Lerner's 1991 book, which Peroomian said is part of a chorus of scientific voices taking down Lerner's theories.

Wright, who taught at the University of California, Los Angeles, published an article refuting Lerner's claims that dark matter doesn't exist or that stars contain less helium than the Big Bang predicted.

"Lerner wants to make helium in stars," Wright said in the article. "This presents a problem because the stars that actually release helium back into the interstellar medium make a lot of heavier elements too."

On dark matter, Wright said the evidence for its existence lies in the orbital motions, bending of light and behavior of gases trapped in clusters of galaxies.

Although scientists cannot see dark matter, they can detect it by measuring how its gravity affects stars and galaxies embedded within it.

Using NASA's Hubble Space Telescope, astronomers learned dark matter forms in much smaller clumps around large and medium-size galaxies than previously known.

Infinity and Our Unbounded Universe

Infinity and Our Unbounded Universe

8.1 Stars Older than the Universe

In 2000, scientists looked to date what they thought was the oldest star in the universe. They made observations via the European Space Agency's (ESA) Hipparcos satellite and estimated that HD140283 — or Methuselah as it's commonly known — was a staggering 16 billion years old.

Such a figure was rather baffling. After all, the age of the universe — determined from observations of the cosmic microwave background — is 13.8 billion years old, so how can a star be older than the universe?

"It was a serious discrepancy," says astronomer Howard Bond of Pennsylvania State University. So with that in mind, Bond and his colleagues set out to discover the truth and test the accuracy of the figure. Their conclusions were just as mind-blowing.

Astronomers began observing Methuselah — named in reference to a biblical patriarch who is said to have died aged 969, making him the longest-lived of all the figures in the Bible — more than 100 years ago. The curious star is located some 190 light-years away from Earth in the

Infinity and Our Unbounded Universe

constellation Libra and it rapidly journeys across the sky at 800,000 mph (1.3 million kilometers per hour).

— Methuselah covers the width of the moon in the night sky every 1,500 years.
— You can't see Methuselah with the naked eye. It can only be seen using a telescope
— It contains just 1/250th of the iron content of our sun

It was clear that the star was old. The metal-poor subgiant is predominantly made of hydrogen and helium and contains very little iron. Such composition meant the star must have come into being when helium and hydrogen dominated the universe and before iron became commonplace (the heavier elements only appeared when massive stars created them in their cores)

But could Methuselah really be more than two billion years older than its environment? Surely that is just not possible. Either the star was older than the universe or the universe was not as "young" as scientists thought it to be. Or maybe the dating was simply all wrong. What was it to be?

A mystery of this magnitude could not be ignored so Bond and his colleagues attempted to unearth the truth by pouring over 11 sets of observations that had been recorded between 2003 and 2011.

These observations had been made by the Fine Guidance Sensors of the Hubble Space Telescope, which noted the positions, distances and energy output of stars. In acquiring parallax, spectroscopy and photometry measurements, the scientists could determine a better sense of age.

Infinity and Our Unbounded Universe

"One of the uncertainties with the age of HD 140283 was the precise distance of the star," Bond said. "It was important to get this right because we can better determine its luminosity and, from that, its age — the brighter the intrinsic luminosity, the younger the star.

"We were looking for the parallax effect, which meant we were viewing the star six months apart to look for the shift in its position due to the orbital motion of the Earth, which tells us the distance."

Bond adds that there were also uncertainties in the theoretical modeling of the stars, such as the exact rates of nuclear reactions in the core and the importance of elements diffusing downwards in the outer layers. So they worked on the idea that leftover helium diffuses deeper into the core, leaving less hydrogen to burn via nuclear fusion. With fuel used faster, the age is lowered.

"Another factor that was important was, of all things, the amount of oxygen in the star," Bond said. HD 140283 had a higher than predicted oxygen-to-iron ratio and, since oxygen was not abundant in the universe for a few million years, it pointed again to a lower age for the star.

As a result of all of this work, Bond and his collaborators estimated HD 140283's age to be 14.46 billion years. It was a significant reduction on the 16 billion previously claimed but it was still more than the age of the universe itself.

In that sense, it didn't clear up the mystery and, on the face of it, simply ensured Methuselah remained a curiosity. But the scientists posed a residual uncertainty of 800 million years, which Bond said made the star's age compatible with the age of the universe. It was a major breakthrough.

Infinity and Our Unbounded Universe

"Like all measured estimates, it is subject to both random and systematic error," said physicist Robert Matthews of Aston University in Birmingham, UK, who was not involved in the study. "The overlap in the error bars gives some indication of the probability of a clash with cosmological age determinations"

"In other words, the best-supported age of the star conflicts with that for the derived age of the universe [as determined by the cosmic microwave background], and the conflict can only be resolved by pushing the error bars to their extreme limits."

Further refinements saw the age of HD 140283 fall a bit more. A 2014 follow-up study, for instance, updated the star's age to 14.27 billion years. "Again, if one includes all sources of uncertainty — both in the observational measurements and the theoretical modeling — the error is about 700 or 800 million years, so there is no conflict because 13.8 billion years lies within the star's error bar," Bond said.

What's more, in May 2021, another group of astronomers revised the best estimates for the age and mass of Methuselah and, having modeled how stars change over time, they found its age to be 12 billion years. It still makes HD 140283 extremely old (the sun, by comparison, is only a kid at 4.6 billion years old) but it puts the age of the star well and truly within the age of the universe. Or does it?

On the one hand, Bond says the efforts to date Methuselah is "an amazing scientific achievement which provides very strong evidence for the Big Bang picture of the universe". By showing similarities between the age of the universe and that of this old nearby star, he says the problem with the age of the oldest stars is far less severe than it was in the 1990s when the stellar ages were approaching 18 billion years or, in one case, 20 billion years. "With the uncertainties of the determinations, the ages are now agreeing," Bond said.

Yet, on the other hand, Matthews believes the problem has not yet been resolved. Astronomers at an international conference of top cosmologists at the Kavli Institute for Theoretical Physics in Santa Barbara, California, in July 2019 were puzzled over studies that suggested different ages for the universe. They were looking at measurements of galaxies that are relatively nearby which suggest the universe is younger by hundreds of millions of years compared to the age determined by the cosmic microwave background.

Far from being 13.8 billion years old, as estimated by the European Planck space telescope's detailed measurements of cosmic radiation in 2013, the universe may be as young as 11.4 billion years. If that is, indeed, the case, then Methuselah is one again older than the universe. The plot, indeed, thickens, but how accurate are these re-estimates proving to be?

One of those behind the studies to date the universe is Nobel laureate Adam Riess of the Space Telescope Science Institute in Baltimore, Maryland.

The conclusions are based on the idea of an expanding universe, as shown in 1929 by Edwin Hubble. This is fundamental to the Big Bang — the understanding that there was once a state of hot denseness that exploded out, stretching space. It indicates a starting point that should be measurable, but fresh findings are suggesting that the expansion rate is around 10% higher than the one suggested by Planck.

Indeed, the Planck team determined that the expansion rate was 67.4 km per second per megaparsec, but more recent measurements taken of the expansion rate of the universe point to values of 73 or 74.

That means there is a difference between the measurement of how fast the universe is expanding today and the predictions of how fast it should be expanding based on the physics of the early universe, Riess said. It's leading to a reassessment of accepted theories while also showing there is still much to learn about dark matter and dark energy, which are thought to be behind this conundrum.

A higher value for the Hubble Constant indicates a shorter age for the universe. A constant of 67.74 km per second per megaparsec would lead to an age of 13.8 billion years, whereas one of 73, or even as high as 77 as some studies have shown, would indicate a universe age no greater than 12.7 billion years.

It's a mismatch that suggests, as stated, that HD 140283 could still be older than the universe. It has also since been superseded by a 2019 study published in the journal Science that proposed a Hubble Constant of 82.4 — suggesting that the universe's age is only 11.4 billion years. Astronomers are hoping the James Webb Space Telescope could shed light on this particular mystery.

Matthews believes the answers lie in greater cosmological refinement. "I suspect that the observational cosmologists have missed something that creates this paradox, rather

Infinity and Our Unbounded Universe

than the stellar astrophysicists," he said, pointing to the measurements of the stars being perhaps more accurate.

"That's not because the cosmologists are in any way sloppier, but because the age determination of the universe is subject to more and arguably trickier observational and theoretical uncertainties than that of stars."

But what could be making the universe potentially appear younger than this particular star?

"There are two options, and the history of science suggests that in such cases the reality is a mix of both," Matthews said. "In this case that would be sources of observational error that haven't been fully understood, plus some gaps in the theory of the dynamics of the universe, such as the strength of dark energy, which has been the prime driver of the cosmic expansion for many billions of years now."

He suggests the possibility that the current "age paradox" reflects time variation in dark energy, and thus a change in the rate of acceleration — a possibility theorists have found might be compatible with ideas about the fundamental nature of gravity, such as the so-called causal set theory. New research into gravitational waves could help to resolve the paradox, Matthews said.

To do this, scientists would look at the ripples in the fabric of space and time created by pairs of dead stars, rather than relying on the cosmic microwave background or the monitoring of nearby objects such as Cepheid variables and supernovae to measure the Hubble Constant — the former resulting in the speed of 67 km per second per megaparsec and the latter in 73.

Infinity and Our Unbounded Universe

Trouble is, measuring gravitational waves is no easy task, given they were only directly detected for the first time in 2015. But according to Stephen Feeney, an astrophysicist at the Flatiron Institute in New York, a breakthrough could be made over the next decade. The idea is to collect data from collisions between pairs of neutron stars using the visible light these events emit to figure out the speed they are moving relative to Earth. It also entails analyzing the resulting gravitational waves for an idea of distance — both of which can combine to give a measurement of the Hubble Constant that should be the most accurate yet.

The mystery of the age of HD 140283 is leading to something bigger and more scientifically complex, altering the understanding of how the universe works.

"The most likely explanations for the paradox are some overlooked observational effect and/or something big missing from our understanding of the dynamics of the cosmic expansion," Matthews said. Precisely what that "something" is, is sure to keep astronomers challenged for some time.

Infinity and Our Unbounded Universe

8.2 Specific Evidence Refuting the Big Bang

The refutation of the Big Bang Theory will have such a major impact on Science, Philosophy, and Religion that we need more evidence of this truth. Here are more specific reasons to support this argument:

1) Light elements: Lithium and Helium

Prediction: Any superhot explosion throughout the universe, like the Big Bang, would have generated a certain small amount of the light element lithium and a large amount of helium.

Observation: Yet as astronomers have observed older and older stars, the amount of lithium observed has gotten less and less, and, in the oldest stars is less than one tenth of the predicted level. The oldest stars near to us have less than half the amount of helium predicted. However, well-understood fusion processes in stars and reactions initiated by cosmic rays have accurately predicted the correct amounts of these and other light elements.

2) Antimatter-matter annihilation

Prediction: Since the intense radiation of the Big Bang would produce matter and antimatter in equal amounts, mutual annulation of particle-antiparticle pairs would reduce the surviving matter density to around 10 -17 protons/cm3.

Infinity and Our Unbounded Universe

Observation: the matter density in the universe is observed to be at least 10^{-7} ions/cm3 more than 10 billion times higher than the Big Bang prediction.

Big Bang fix to prediction: To try to fix this well-known vast gap, Big Bang theorists have proposed some unknown asymmetry between matter and antimatter which would lead to more production of matter. This has never been observed in laboratory experiments. A consequence of this predicted imbalance is the decays of the proton, initially predicted to decay with a lifetime of 10 to the 30th years. Large scale experiments have contradicted this prediction was well, with no evidence of decay at all.

3) Surface-Brightness

Prediction: In any expanding universe, an optical illusion makes objects at high redshift appear larger and dimmer, so their surface brightness—the ratio of apparent brightness to apparent area—declines sharply with redshift.

Observation: Based on observations of thousands of galaxies, surface brightness is completely constant with distance, as expected in a universe that is NOT expanding.

Big Bang fix to Prediction: After observations showed that the surface brightness dimming did not occur, Big Bang theorists hypothesized that galaxies were much smaller in the distant past and have grown greatly. But observations have contradicted this fix as well, showing that there have not been enough galaxy mergers for the growth rates needed. In addition, the ultra-small galaxies hypothesized would have to have more mass in stars than total mass, an obvious impossibility.

4) Too Large Structures

Prediction: In the Big Bang theory, the universe is supposed to start off completely smooth and homogenous. Structure starts small and grows over time

Observation: As telescopes have peered farther into space, huger and huger structures of galaxies have been discovered, which are too large to have been formed in the time since the Big Bang.

5) Cosmic Microwave Background Radiation (CMB) and its Anisotropies

Prediction (Initial): The CMB is a smooth relic of the initial radiation of the Big Bang.

Observation: The CMB is smooth on such large scales that in a Big Bang there would be too little time for regions that we now see in different parts of the sky to reach equilibrium with each other, or even to receive energy from each other at the speed of light.

Big Bang fix to prediction: An unknown force, dubbed "inflation" generated an exponential phase of the Big Bang that blew up the universe so rapidly that all asymmetries were smoothed away.

Infinity and Our Unbounded Universe

Additional observations: The actual very small anisotropies in the CMB were much smaller than those predicted by Big Bang theorists and additional fixes had to be added to the theory each time the observations became more precise, so that at present seven free variables—the density of dark matter, of ordinary matter, of dark energy and four additional fitting parameters—are needed to fit the observations. They still badly fail with some of the largest-scale anisotropies.

The latest crisis: Based on the data from the Planck satellite, the best fit to the CMB predicts a Hubble constant (the ratio of redshift to distance) in conflict with observations based on Supernovae. The best fits imply a curved universe, in conflict with the predictions of inflation for a flat universe. And they predict a density of dark matter far greater than any measurements derived from the motion of galaxies.

In contrast to the multiple contradictions of the Big Bang theory of the CMB with its "ultra-precise" but wrong predictions, non-Big Bang processes provide a better explanation. The energy that was released in producing the observed helium in the universe equal the energy in the CMB. Any radiation become isotropized if it travels in a medium that scatters it. There is abundant observational evidence that microwave-frequency radiation is scattered in the intergalactic medium.

6) Dark Matter

Prediction: The Big Bang theory requires the existence of dark matter—mysterious particles that have never been observed in the laboratory, despite huge experiments to find them.

Observation: Multiple lines of evidence, especially observations of the motions of galaxies, show that this dark matter does not exist. Extremely sensitive experiments on earth have failed to detect dark matter particles. In addition, dark matter, if it existed would create a viscosity effect on galaxies that would prevent the existence of the many long-lived groups of galaxies that are observed.

The response of most cosmologists to this growing body of evidence has, unfortunately, not been to decide the Big Bang theory has been falsified, but to add new "parameters" and hypotheses, like dark energy. The theory is now far more complex and speculative than the Ptolemaic epicycles that were destroyed by the Scientific Revolution. Each contradiction with observation is taken as a mere "anomaly" that does not undermine the theory as a whole. Strong peer pressure is applied against many of those who question the theory.

"It's as if researchers are saying 'I can see the Emperor's elbow through his New Clothes,' 'I can see the Emperor's knee though his New Clothes' and so on," says Lerner. "It is time to say: 'The Emperor is not wearing any clothes.' This theory has no correct predictions."

To replace the Big Bang, other researchers have elaborated, in peer-reviewed publications, alternative explanations of the generation of light elements and of the energy in the CBR by ordinary stars, and of the development of large-scale structures through the interaction of gravity and electromagnetic processes. "No one would claim that all the problems in cosmology have been resolved," agrees Lerner, "but the evidence is consistent with an evolving, but non-expanding universe, which had no beginning in time and no Big Bang."

7) Galaxies exist which are too old to have been created in the Big Bang.

To everyone who sees them, the new James Webb Space Telescope (JWST) images of the cosmos are beautifully awe-inspiring. But to most professional astronomers and cosmologists, they are also extremely surprising—not at all what was predicted by theory. In the flood of technical astronomical papers published online since July 12, the authors report again and again that the images show surprisingly many galaxies, galaxies that are surprisingly smooth, surprisingly small and surprisingly old. Lots of surprises, and not necessarily pleasant ones. One paper's title begins with the candid exclamation: "Panic!"

Why do the JWST's images inspire panic among cosmologists? And what theory's predictions are they contradicting? The papers don't actually say. The truth that these papers don't report is that the hypothesis that the JWST's images are blatantly and repeatedly contradicting is the Big Bang Hypothesis that the universe began 14 billion years ago in an incredibly hot, dense state and has been expanding ever since. Since that hypothesis has been defended for decades as unquestionable truth by the vast majority of cosmological theorists, the

Infinity and Our Unbounded Universe

new data is causing these theorists to panic. "Right now I find myself lying awake at three in the morning," says Alison Kirkpatrick, an astronomer at the University of Kansas in Lawrence, "and wondering if everything I've done is wrong."

8) The Universe Expanded too fast

If the Big Bang really occurred 13.7 years ago we would expect it to have a radius of about the same number of light years or less. This is because the universe's matter can't expand as fast as the speed of light.

However, the current size of the Universe (the limits of our observations) is 94 billion light years in diameter.

This means that when the Big Bang happened the Universe would have had to expand faster than the speed of light for it to become this large.

We don't have any explanations or scientific justification to say that the early Universe somehow expanded at a speed greater than the speed of light.

Infinity and Our Unbounded Universe

Infinity and Our Unbounded Universe

8.3 Dark Matter Doesn't Exist?

Lets look at one of the declarations in the previous chapter about Dark Matter in more detail to better understand why the Big Bang probably never happened which is more in line with the idea that we actually live in an infinite and steady state Universe.

Dark Matter May Not Exist: Some Physicists Favor of a New Theory of Gravity

Dark matter was proposed to explain why stars at a galaxy's far edge were able to move much faster than predicted with Newton. An alternative theory of gravity might be a better explanation.

Using Newton's laws of physics, we can model the motions of planets in the Solar System quite accurately. However, in the early 1970s, scientists discovered that this didn't work for disc galaxies – stars at their outer edges, far from the gravitational force of all the matter at their center – were moving much faster than predicted by Newton's theory.

Infinity and Our Unbounded Universe

As a result, physicists proposed that an invisible substance called "dark matter" was providing extra gravitational pull, causing the stars to speed up – a theory that's become widely accepted. It is suggested that observations across a vast range of scales are much better explained in an alternative theory of gravity called Milgromian dynamics or Mond – requiring no invisible matter. It was first proposed by Israeli physicist Mordehai Milgrom in 1982.

Mond's primary postulate is that when gravity becomes very weak, as it does near the edge of galaxies, it starts behaving differently from Newtonian physics. In this way, it is possible to explain why stars, planets, and gas in the outskirts of over 150 galaxies rotate faster than expected based on just their visible mass. However, Mond doesn't merely *explain* such rotation curves, in many cases, it *predicts* them.

Philosophers of science have argued that this power of prediction makes Mond superior to the standard cosmological model, which proposes there is more dark matter in the universe than visible matter. This is because, according to this model, galaxies have a highly uncertain amount of dark matter that depends on details of how the galaxy formed – which we don't always know. This makes it impossible to predict how quickly galaxies should rotate. But such predictions are routinely made with Mond, and so far these have been confirmed.

Imagine that we know the distribution of visible mass in a galaxy but do not yet know its rotation speed. In the standard cosmological model, it would only be possible to say with some confidence that the rotation speed will come out between 100km/s and 300km/s on the outskirts. Mond makes a more definite prediction that the rotation speed must be in the range 180-190km/s.

If observations later reveal a rotation speed of 188km/s, then this is consistent with both theories – but clearly, Mond is preferred. This is a modern version of Occam's razor – that the simplest solution is preferable to more complex ones, in this case that we should explain observations with as few "free parameters" as possible.

Free parameters are constants – certain numbers that we must plug into equations to make them work. But they are not given by the theory itself – there's no reason they should have any particular value – so we have to measure them observationally. An example is the gravitation constant, G, in Newton's gravity theory or the amount of dark matter in galaxies within the standard cosmological model.

We introduced a concept known as "theoretical flexibility" to capture the underlying idea of Occam's razor that a theory with more free parameters is consistent with a wider range of data – making it more complex. In our review, we used this concept when testing the standard cosmological model and Mond against various astronomical observations, such as the rotation of galaxies and the motions within galaxy clusters.

Each time, we gave a theoretical flexibility score between –2 and +2. A score of –2 indicates that a model makes a clear, precise prediction without peeking at the data. Conversely, +2 implies "anything goes" – theorists would have been able to fit almost any plausible observational result (because there are so many free parameters). We also rated how well each model matches the observations, with +2 indicating excellent agreement and –2 reserved for observations that clearly show the theory is wrong. We then subtract the theoretical flexibility score from that for the agreement with observations, since matching the data well is good – but being able to fit anything is bad.

A good theory would make clear predictions that are later confirmed, ideally getting a combined score of +4 in many different tests (+2 -(-2) = +4). A bad theory would get a score between 0 and -4 (-2 -(+2)= -4). Precise predictions would fail in this case – these are unlikely to work with the wrong physics.

We found an average score for the standard cosmological model of –0.25 across 32 tests, while Mond achieved an average of +1.69 across 29 tests. The scores for each theory in many different tests are shown in figures 1 and 2 below for the standard cosmological model and Mond, respectively.

Infinity and Our Unbounded Universe

	Clear prior expectation	Not predicted, but follows from theory	Auxiliary assumptions needed, but these have little effect	Auxiliary assumptions needed, but these have a discernible effect	Auxiliary assumptions allow theory to fit any plausible data
Excellent agreement	● Gravitational waves travel at c ● Expansion history at $z \gtrsim 0.2$	● Einstein ring radii		○ CMB anisotropies	● MW escape velocity curve ● MW-M31 timing argument ● Galaxy cluster internal dynamics ● Galaxy two-point correlation function
Works well	● Big Bang nucleosynthesis ● Offset between X-ray and lensing in Bullet Cluster			● Hickson Compact Group abundance	
Plausibly works	● Weak lensing correlation function		● Galaxy cluster mass function at low redshift		● Weak lensing by galaxies ● HSB disc galaxy RCs
Some tension	● Number of spiral arms in disc galaxies ● External field effect			● Prevalence of thin disc galaxies ● Weakly barred M33	● LSB disc galaxy RCs ● Gas-rich galaxy RCs ● Elliptical galaxy RCs ● Spheroidal galaxy σ_{LOS} ● Galaxy group σ_{LOS}
Strong disagreement	● No distinct tidal dwarf mass-radius relation ● Local Group satellite planes ● El Gordo formation ● KBC void ● Local Hubble diagram slope and curvature	● Galaxy bar pattern speeds ● RV of NGC 3109 association	● Tidal limit to radii of MW satellites ● Bar fraction in disc galaxies		

Figure 1. Comparison of the standard cosmological model with observations based on how well the data matches the theory (improving bottom to top) and how much flexibility it had in the fit (rising left to right). The hollow circle is not counted in our assessment, as that data was used to set free parameters.

Infinity and Our Unbounded Universe

	Clear prior expectation	Not predicted, but follows from theory	Auxiliary assumptions needed, but these have little effect	Auxiliary assumptions needed, but these have a discernible effect	Auxiliary assumptions allow theory to fit any plausible data
Excellent agreement	● Gravitational waves travel at c ● Expansion history at $z \gtrsim 0.2$	● Einstein ring radii		⊙ CMB anisotropies	● MW escape velocity curve ● MW-M31 timing argument ● Galaxy cluster internal dynamics ● Galaxy two-point correlation function
Works well	● Big Bang nucleosynthesis ● Offset between X-ray and lensing in Bullet Cluster			● Hickson Compact Group abundance	
Plausibly works	● Weak lensing correlation function		● Galaxy cluster mass function at low redshift		● Weak lensing by galaxies ● HSB disc galaxy RCs
Some tension	● Number of spiral arms in disc galaxies ● External field effect			● Prevalence of thin disc galaxies ● Weakly barred M33	● LSB disc galaxy RCs ● Gas-rich galaxy RCs ● Elliptical galaxy RCs ● Spheroidal galaxy σ_{LOS} ● Galaxy group σ_{LOS}
Strong disagreement	● No distinct tidal dwarf mass-radius relation ● Local Group satellite planes ● El Gordo formation ● KBC void ● Local Hubble diagram slope and curvature	● Galaxy bar pattern speeds ● RV of NGC 3109 association	● Tidal limit to radii of MW satellites ● Bar fraction in disc galaxies		

Figure 2. Similar to Figure 1, but for Mond with hypothetical particles that only interact via gravity called sterile neutrinos. Notice the lack of clear falsifications.

It is immediately apparent that no major problems were identified for Mond, which at least plausibly agrees with all the data (notice that the bottom two rows denoting falsifications are blank in figure 2).

The problems with dark matter

One of the most striking failures of the standard cosmological model relates to "galaxy bars" – rod-shaped bright regions made of stars – that spiral galaxies often have in their central regions (see lead image). The bars rotate over time. If galaxies were embedded in massive halos of dark matter, their bars would slow down. However, most, if not all, observed galaxy bars are fast.

This falsifies the standard cosmological model with very high confidence.

Another problem is that the original models that suggested galaxies have dark matter halos made a big mistake – they assumed that the dark matter particles provided gravity to the matter around it, but were not affected by the gravitational pull of the normal matter. This simplified the calculations, but it doesn't reflect reality. When this was taken into account in subsequent simulations it was clear that dark matter halos around galaxies do not reliably explain their properties.

There are many other failures of the standard cosmological model that we investigated in our review, with Mond often able to naturally explain the observations. The reason the standard cosmological model is nevertheless so popular could be down to computational mistakes or limited knowledge about its failures, some of which were discovered quite recently. It could also be due to people's reluctance to tweak a gravity theory that has been so successful in many other areas of physics.

The huge lead of Mond over the standard cosmological model in our study led us to conclude that Mond is strongly favored by the available observations. While we do not claim that Mond is perfect, we still think it gets the big picture correct – galaxies really do lack dark matter.

Infinity and Our Unbounded Universe

Infinity and Our Unbounded Universe

9.0 Implications of an Infinite Universe

Our entire cosmology of the Universe is currently based on the Big Bang Theory. Many scientists even try to come up with explanations for the discrepancies which are cropping up which would tend to disprove the Big Bang really happened.

If our Universe is truly unbounded in space and possibly with an infinite history this has incredible implications for our lives and thinking.

Questions which occur to me and their elaborations are as follows:

1) If the Universe was not created does God Really exist?

I believe that God is the life force of the Universe which is present everywhere. So it doesn't matter whether there was a Big Bang or the Universe is steady state since God is infinite anyway and beyond our understanding.

2) Was this steady state Universe created at some previous point in time or has it always existed?

This is a really interesting thought. Has the Universe always existed or was it created at some point in time too?

Ancient stars we have found and galaxies which we believe were the first to be created would have a definite starting point in time.

However, we don't know if previous astronomical structures existed and are no longer in existence.

Or how about the idea that if the Universe is infinite in size and maybe other stars and galaxies existed in times

Infinity and Our Unbounded Universe

previously and which no longer exist or are beyond the visual perception our greatest telescopes have of the Universe.

3) If there are an almost infinite number of galaxies and stars do Earth and ourselves have duplicates somewhere else in the Universe?

If this is true then logically there must be duplicates of Earth and each of us Individuals somewhere in the Universe.

This leads to one of my favorite metaphysical speculations which is if there are duplicates of me out there somewhere then can we be in telepathic sync with each other?

I sometimes visualize this when going to sleep to see what types of impressions I might receive from my duplicates across the Universe. An interesting exercise.

4) Currently scientists say that two trillion galaxies exist in our Universe. Are there really an infinite number of galaxies which also means an infinite variety of life?

This is a question we don't know the answer to and may never know the answer to because the truth is beyond the limits of our perceptions.

5) Could there be different fundamental constants in different parts of the Universe?

This has been a speculation by a number of physicists for a number of years. The idea is that gravity or the speed of light for example might be different at other locations in the Universe.

Infinity and Our Unbounded Universe

Currently there is no evidence that this is the case.

6) Could alternative dimensions actually be different places in our Universe and not in some type of separate dimensional realm?

I have done a lot of research on alternative dimensions and in my book titled "Stories of Parallel Dimensions" I found a lot of first person accounts of traveling to other dimensions and back which seemed valid to me.

So I believe in the concept of alternative dimensions existing. Who is to say if these dimensions are real or just very different locations in our Universe which are beyond our perception?

Maybe those people who crossed over went through some types of time and distance gateways to get there and those locations are really just part of our Universe?

Infinity and Our Unbounded Universe

Infinity and Our Unbounded Universe

10.0   Summary

Infinity is an amazing concept which underpins much of the mathematics used in our world today for Engineering, Physics, Chemistry, and more.

Infinity also has implications for our world as a whole. I don't know if the Universe is actually infinite in size and age because we can't see it all with our best instruments.

However, the evidence is becoming overwhelming that there was never a Big Bang which created the Universe. So the Universe is likely steady state and is at least much older and larger than we can determine.

We can think of but can't actually visualize the concept of infinity. An interesting contradiction of our thinking processes.

I hope this book stirred some questions and new thoughts in your minds too.

All the Best,

Martin K. Ettington
November 2022

Infinity and Our Unbounded Universe

11.0 Bibliography

Ettington, M. K. (2021). *Stories of Parallel Dimensions.*

https://moosmosis.org/2022/06/14/calculus-infinite-limits-and-limits-at-infinity-with-explanations-practice-questions-and-answers-ap-calculus-calculus-101-math/. (2022). Retrieved from Infinity Explanations.

https://scitechdaily.com/dark-matter-may-not-exist-these-physicists-favor-of-a-new-theory-of-gravity/. (2022). Retrieved from Dark Matter May Not Exist.

https://www.space.com/how-can-a-star-be-older-than-the-universe.html. (2022). Retrieved from How Can a Star be Older than the Universe.

www.ingramcontent.com/pod-product-compliance
Lightning Source LLC
Chambersburg PA
CBHW071147240526
45465CB00024BA/1821